Majestic
MONGOLIA

Dr. Diana Prince

AuthorHouse™
1663 Liberty Drive
Bloomington, IN 47403
www.authorhouse.com
Phone: 833-262-8899

Because of the dynamic nature of the Internet, any web addresses or links contained in this book may have changed since publication and may no longer be valid. The views expressed in this work are solely those of the author and do not necessarily reflect the views of the publisher, and the publisher hereby disclaims any responsibility for them.

This book is printed on acid-free paper.

ISBN: 978-1-6655-7904-9 (sc)
ISBN: 978-1-6655-7905-6 (hc)
ISBN: 978-1-6655-7903-2 (e)

Library of Congress Control Number: 2022923803

Print information available on the last page.

Published by AuthorHouse   01/10/2023

authorHOUSE®

Table of Contents

Introduction

Nearly half of all Mongolians follow the nomadic existence of their ancestors. Yaks, camels, sheep and horses occupy the magnificent landscape.

Geographically speaking, Mongolia is perhaps the most unusual place in the world to find a thriving democracy. It is bordered, on all sides, by two aggressive regimes—Russia in the north and China in the south.

Today, Mongolia is a free republic. However, the two political giants—China and Russia—each formerly controlled what is now a free and democratic Mongolia. This land is part of the great Mongolian Empire built by Genghis Khan. Genghis Khan was a fierce warrior who united many ancient tribes into the Mongolian empire. When Khan swept into what is now Mongolia, the vast empire was the greatest expanse of territory ever conquered by any individual in the history of the world.

Chinese rule in this area had dated from 1206, until Genghis Khan, the powerful Mongol ruler and warlord seized control of what is now Mongolia, and ruled with an iron hand.

Kublai Khan, the grandson of the great Genghis Khan, later conquered China. This was the era of the Yuan dynasty. When this happened, many of the Mongols left China and made their home in Mongolia.

Religion

In the late 1500's, Tibetan Buddhism was introduced to Mongolia. Most of this was due to the followers and descendants of Genghis Khan.

In its early history, Mongolia embraced Buddhism. This period lasted over a two-thousand year period. The original Buddhist monks came to Mongolia from Nepal. They were welcomed by both the Rowan Khaganate and the Hunnu Xianbe Empires.

Today, over half the people of Tibet still identify themselves as Buddhists. About forty percent of the people, however, do not identify with any specific religion.

Islam, Christianity and other religions combined, comprise less than 10 percent of religious preference among Mongolians.

Today, there is no religious persecution in Mongolia. This country, with a varied religious past, and former religious strictures under both Russia and China, now has a free and open tolerance of religious differences.

Prior to this century, Mongolia was a nomadic, feudal state under China during the Manchu Dynasty beginning in 1691. This lasted for more than 200 years. In a ten-year war beginning in 1911, the country emerged from Chinese control. Mongolia declared its independence in 1921.

The Soviet Union had helped Mongolia assert its independence from China. In the process, Mongolia soon came under the control of the Soviet Union, itself.

Almost immediately after ending its long association with China, Mongolia was subjected to Russian rule, and had exchanged one political master for another.

Mongolia operated for almost 70 years under Russian control, and subjected to their Communist regime. The Mongolian People's Republic was a socialist entity.

The New Democracy

In 1989, anti-communist sentiment had weakened Soviet power and control over Mongolia. In May of 1990, Mongolia transitioned from Soviet control to a democracy in what was called the "Peaceful Revolution". In 1992, Mongolia introduced a new constitution. Today, Mongolia is, unexpectedly, a thriving democracy.

There is an irony in this clearly unlikely transition to democracy, especially given Mongolia's long history. The change to a democratic society is even more profound, given that the country is bordered entirely in the north by Russia, and entirely in the south by China. Facing such restrictive and controlling regimes both to the north and the south will be challenging as Mongolia emerges onto the world stage. It is a daunting experiment for this new democracy.

The significant changes, over a relatively short period of time, have brought Mongolia a major period of adjustments, changes and challenges.

In March of 2022, Mongolia presented a written declaration to the United Nations voicing its position on weapons of war. In its contents it stated that *"Mongolia does not own, possess, or control nuclear weapons, and has never done so, and it does not host any other state's nuclear weapons on its territory."*

The area known as "Inner Mongolia" refers to a part of Mongolia which is still part of China, not currently a part of the Republic of Mongolia.

Mongolia has experienced many changes since becoming a democracy. Today Mongolia is a partner nation with NATO. In 1997, the country became a member of the World Trade Organization.

Today, Russia has an established embassy at Ulaanbaatar, and conversely, Mongolia maintains an embassy in Russia at Moscow.

This book examines the specific areas of Mongolia by region. Mongolia extends from the tundra in Siberia to the Gobi Desert. Almost half of Mongolia has rich green hills and mountains. The Altai Mountain chain dominates Western Mongolia. In this region, there are also hundreds of glaciers. Some areas are covered in snow all year long.

This book presents a view of Mongolia in this century. It is a land of challenges, and a place of rugged beauty.

Above all, I dedicate this book to the remarkable people of Mongolia whom I have had the privilege of meeting.

CHAPTER 1

Northern Mongolia

The Khosvol Lake Region

Mongolia's largest and deepest lake is Lake Khosvol located in the northernmost part of Mongolia. Lake Khosvol is nicknamed the "Blue Pearl" of Mongolia. Lake Khosvol holds about seventy percent of the fresh water in all of Mongolia. It lies 5,000 feet above sea level, and has a depth of over 800 feet. It is located near Mongolia's northern border with Russia. It stretches 85 miles in length, and is over twenty-two miles wide. The lake, formed about three million years ago, is one of the world's oldest lakes. It is also one of the top freshwater sources in the world.

In the broad blue waters of Lake Khosvol, there are several large islands. Two of the largest are Modon Khui and Khadan Khui. Lush forests of pines rise up in the foothills above the lake.

The only lake in Mongolia larger than Lake Khosvol is Lake Uvs. Lake Uvs has a greater surface area by length and width. However, Khosvol Lake is the largest by volume, with a greater depth.

In winter, Lake Khosvol freezes over so completely that horses have been observed pulling sleds and passengers across the ice. Cars also have taken their chances by driving on the lake when it is solidly frozen over in winter.

Lake Khosvol is also called the "sister lake" of nearby Lake Baikal. Both lakes originated during the same geological period. While Lake Khosvol is in the northernmost part of Mongolia, Lake Baikal lies directly across the Russian border in the area known as Siberia.

Lake Baikal is popular for climbers and hikers. It is the largest and deepest freshwater lake in the world. It lies at an altitude of 1500 feet. It is also the world's oldest lake. It is said to have originated over 20 million years ago.

It is one of the main sports, hiking and vacation areas for Mongolians who live near the border in northern Mongolia.

Horses have always been vital to Mongolian life. There are more horses in this region than in any part of Mongolia. Horses play a major role in Mongolia's culture. Horses were critical centuries ago when Genghis Khan first waged his war to subdue this area. The horse continues to be vital in the challenging rural environment.

Today there are over three million domesticated horses in Mongolia. The horses bred for this severe and demanding environment are very rugged. They tend to be shorter but more stocky and muscular, when compared to European horse breeds. These horses are vital to this environment where herding is central to rural life and communities. Riders generally use hard saddles with wooden components. Riders are also trained to work with horses, even as children.

Wildlife Near Khosvol Lake

Wild horses were once plentiful in the nearby hills. Falcons and eagles still fly across the wide expanse of sky. In the hills above the placid lake, a forest of pines rises against the sky.

The rare Siberian musk deer also thrives in this region. However, it occupies high altitudes and is rarely seen. There are also over 150 species of birds in this region.

Dinosaur fossils, found in this area, date back 70 million years.

This is the only place in the world where reindeer are domesticated. This is an ethnic practice by a tribe named the Tsaatan. They originally migrated here from Turkey. They are known specifically for their skills of herding reindeer. They practice a religion of Shamanism, which involves the worship of nature.

The Town of Hatgal

Hatgal, at the southern point of Lake Khovsgol, was founded in the early 1700's. The scenic village has a rustic wooden bridge that spans the Eg River. It was originally a small outpost at the border with Russia.

In the early 1900's, it had established active trading with Russia. By the end of the 1990's, many people had migrated here, and the population had reached over 7,000 people.

However, an economic downturn occurred when local industries left, in particular the large wool processing factory. By the turn of the century, over half the town had left.

Later when Mongolian Airlines made Hatgal an itinerary stop on its route, it restored jobs and tourism.

Today it is also known as a host for the largest Naadam Festival in Mongolia.

The Naadam Festival at Hatgal

The National Festival of Naadam is held at Hatgal near Lake Khosvol each summer. The Naadam playoffs have also been held at Ulaanbaatar in central Mongolia.

The three major sports in Mongolia are wrestling, horse racing and archery competitions. The most popular of these is wrestling, a sport which dates back to ancient warriors in Mongolia over 7,000 years ago. It is officially called "Bokh".

The sport of wrestling is now under the auspices of the Mongolia Wrestling Association. This commission organizes the local competitions. It sets rules for the competition at the local level, and determines which regional winners will move on to the final national competitions. Those selected will compete in the national Naadam Festival.

Naadam wrestling is characterized by one important rule. The participant will lose the competition if he touches the ground with any part of his body except his foot.

The Naadam wrestling event is a two-day competition. It is usually held on July 11th. In each competition, about 500 wrestlers compete with one another. After a series of competing rounds, the weaker competitors are gradually eliminated. Those who persevere in smaller village competitions will advance to the higher national competitions.

Over a period of days, the festivities highlight events with titles such as the "Elephant Round" and the "Falcon Round", each with their specific emphasis.

These events are so prestigious that the winning finalists are presented their awards by the President of Mongolia.

Driftwood near Lake Khovsgol

Cloudy Day at the Shore

Cattle relaxing in Lake Khovsvol

Ger in west Khovsgol

Ger Village in Late Afternoon

Ger Village near Forest

View of Mountains Overlooking Lake Khovsgol

Old stone pier at Lake Khovsgol

Late August at Lake Khovsgol

Brilliant Cloud Display Over Mountains

Lake Reflections

Peaceful morning at the Lake

Clouds gather over coastal village

Tourist Boat at Lake Khovsgol

Picnic at the Lake

Old Patrol Boat

Sunny Morning in Town

Late Afternoon at the Ger Camp

Sunset Walk on the Lake Shore

Campfire at Water's Edge

Friends gather at Sun Down

Village of Gers near the Forest

Ger overlooking the Lake

Gers amid the Trees

Cattle Gathering at the Lake

Rustic ger on the grassy plain.

Girl on small farm.

Ger at small farm.

Bull Resting on Farm.

Woman Working on Farm

Young Girl on Farm

Cattle Resting on Farm

Milking on the Farm

On the Way to Visit a Family Inside a Ger

This Man and His Wife Await Guests

The Ger Owner Wears Traditional Clothes
that are typical in this Region.

The Wife is Welcoming Visitors in Their Home.

The Woman Prepares Lunch with Mongolian Specialties

The Wife Prepares the Traditional Mongolian Dinner
using the Small Iron Stove in the middle of the Ger.

Local Delicacies come from the Fresh Produce on a Nearby Farm.

The Couple Describe their Life in the Lake Khovsgol Region.

Traditional Clothing and Furniture inside the Ger.

Closeup of Traditional Headwear and
Coat in this Often Cold Climate.

Colorful Handmade Items Lend a Warmth to the Home.

Mongolian Flag at the Nadaam Sports Wrestling Competition

Contestants in the Nadaam Wrestling Competition

Young Boys Watch the Wrestling Teams Compete
in the Nadaam Wrestling Championship.

Teams Contend for the Nadaam Wrestling Championship

CHAPTER 2
Southern Mongolia

Gobi Desert Region

The Gobi Desert lies in the southern part of Mongolia. The "Flaming Cliffs" area is known for its spectacular red rock towering above the desert floor. This region is also known for the research of Roy Chapman Andrews in 1920. This American paleontologist discovered dinosaur eggs in this region, formerly known by the name of "Bayanzag".

Dinosaur skeletons have been found in this region of the Gobi Desert, as well as cave paintings.

The soil in this region turns a rich reddish color at sunset. Vultures rise up over the towering mountains in the Yolyn Am Canyon. Today, new fossil hunters search this area for new discoveries.

The vast Gobi Desert occupies nearly one-third of Mongolia. In the Gobi, ever-shifting sand dunes lie in a panorama of vast beauty. The vast Gobi Desert comprises over twenty percent of the earth.

There is an unearthly calm that pervades the sky over the Gobi Desert. There is an expanse of sky that goes on interminably, and envelopes both human beings and other creatures with a keen sense of the life force which endures and flourishes here. There is a calm which renders all boundaries useless.

Here, in this isolated, beautiful land, Nomads let their animals roam freely, and systematically gather them together when it is time to move to new grazing lands. Vast herds of camels, goats and cattle roam undisturbed in a land with no fences or boundaries.

And in all of it, there is something solid and pervading, as if the heartbeat of this world beats to the primitive roots in all of us.

Most people in this region live in "yurts", which are more commonly called "gers" in Mongolia. These desert homes are round structures constructed with cloth and canvass over wooden frames that can be easily moved from place to place.

For the nomads, these gers are very mobile. They seasonally disassemble their homes, and carry them with them, to set their gers up in a new location. Often the reason is to find new and fresh grazing land for their animals.

In less than four hours, a ger can be leveled to the floor, and totally reassembled again in the new location. The rugged canvas walls of the gers are resilient and can withstand heavy rainfall.

The interior of the gers is usually furnished with simple frame furniture, as well as elaborate blankets with homemade needlework. These add a comfortable and warm feeling of "home".

Even in heavy downpours in the desert, the gers are resilient. They are warm and insulated by the cloth that pads the inner layer of the canvas.

Each ger is also well heated by a small central iron stove which occupies the center of each ger. It is a unique experience to sleep, ensconced in warm woolen blankets, and listen to the rain and howling winds outside.

Staying in yurts on the barren Gobi plateaus is a one-of-a-kind experience. In the early evening, the animals spread over the plains under the reddest of sunsets, not in small herds but in waves of goats, sheep, yaks, horse and cattle mixed together and moving like one body against the horizon. In the morning the herders head out to find their animals, and begin anew the brewing, baking, sewing, rope-making, and care of their children. There are no wasted moments, no leftover time.

Every few months they pack up their belongings, dissemble their tents and make their seasonal moves, and begin again.

The southern Gobi Desert is a virtual treasure trove of dinosaur discoveries. Speculation is that millions of years ago, this area was an inland sea. Sand slides, a changing terrain over millennia, and other natural occurrences have somehow, against all odds, managed to perfectly preserve the bone specimens. Most of the dinosaurs were determined by paleontologists, to have occupied this region about 80 million years old, roughly during the Late Cretaceous Period.

Fast forwarding to a time of human occupation on this arid and rugged plateau, there are petroglyphs depicting animals and hunting scenes dating back 6,000 years ago. The natural isolation had allowed these artifacts to rest undisturbed over time. It appears that scientists are only at the tip of finding the wealth of history and artifacts that lie buried here.

On the desolate plains of Ukhaa Tolgod, when the one main road ends, vehicles forge their own roads, etching rugged tracks over unexplored new ground. Tire marks leave rutted indentations in the earth that crisscross endlessly. A GPS will do you no good in this vast and unending desert.

A particular area of interest is the area called the Flaming Cliffs, which was point zero for the field operations of paleontologist Ray Chapman Andrews and his crew of researchers in the 1920's. In this place, fossil beds dating to 80 million years ago were first uncovered. The Flaming Cliffs are named for their deep red rock towering above the desert floor in dramatic shapes. A rich cache of dinosaur eggs was extracted by Chapman's team. One later specimen was a dinosaur egg, with part of the shell peeled away. It revealed the small embryo of a baby dinosaur intact inside the shell. The scientists were speechless about their finds.

In 1923, Roy Chapman Andrews, the most prominent paleontologist in this area was to make a new groundbreaking discovery. When he unearthed an oviraptor dating to the Late Cretaceous period, it turned dinosaur research on its head, questioning our perceptions of how these desert giants functioned in their families and interacted within their species group.

For several years afterwards, before Mongolia rejected Soviet rule in 1990, Americans were not eager to do research in this region. This was partly due to political considerations, but also to the difficulty of functioning in this place of unmapped and remote outposts. In 1993, Americans returned to this part of the world, and to the rich deposits of the Flaming Cliffs area. Here researchers discovered a large pit of 60 dinosaurs in close proximity. These specimens were in an extraordinarily well preserved condition. This alone fueled the quest to do further research.

Camels in Late Afternoon

Sun Setting in South Gobi Desert

Gers at Sunset

Learning to Explore

Sunset over the Gobi

Holding a Conference

Heading Home

Magnificent Desert Sky

The Living Desert

Prehistoric Engraving in Stone

The Author Riding a Camel in the South Gobi Desert

Baby Camel with Mother

Baby Camels Following Mother

Morning on the Gobi

Getting Directions

Climber overlooking the "Flaming Cliffs" at Bayanzag

The "Flaming Cliffs" of Bayanzag

Early Morning Walk in the Desert

Sunset on the Gobi Desert

Heading East on the Gobi

Morning Meeting on the Gobi

Getting There

Setting Up Camp

Heading East

Amazing Variety on the Desert

Holding a Conference

Desert Sunset on the Gobi

Taking a Break on the Gobi

Among Friends

Taking a Break

Young Girl in front of her Family Ger in the Gobi Desert

Playing Inside the Ger

Mother Working inside the Ger

Young Boy Helping with the Family Dinner.

Brother and Sister Playing with a Camel on the Gobi

Cattle Head Home under a Perfect Sunset on the Gobi

"Looks like rain."

CHAPTER 3

Western Mongolia

Altai Mountain Region

Western Mongolia is the place where Mongolia, China and Russia converge in a region called the Altai Mountain area. Reaching an elevation near 15,000 feet, Belukha Mountain is the highest peak in this region. It was formed by tectonic forces nearly 400 million years ago.

The rugged and mountainous area of the Altai Mountain Region has a rich cultural legacy dating to prehistoric Mongolia. There is evidence of inhabitants here as early as the Bronze Age. Their cave drawings and petroglyphs have been preserved here.

In this region, cave paintings have been found which date back over 20,000 years. In one former Mongolian outpost, now under Russian control, ancient stone figurines date back thousands of years. A mummy dating back over 4,000 years was also unearthed in the Altai Region of Mongolia. Vivid cave paintings today preserve the evidence of thriving communities that existed in this region for thousands of years.

The Altai Mountain region is also known for its famous "eagle hunters". Each year, the Golden Eagle Festival is held at Bayan-Olgii in this region.

The Kazakhs are famous for their eagle hunting skills. They are the largest of Mongolia's many ethnic groups. Beginning in September each year, these eagle hunters hold elaborate festivals, which feature their keen skills. These competitions last for two months.

This eagle hunting tradition dates back for over a thousand years. It is still practiced by local hunters on horseback. These skilled hunters, who used trained eagles to help catch their prey, were even mentioned centuries ago, by Marco Polo. The tradition is also popular in many other parts of Mongolia, where eagle hunters also compete each year in local competitions.

Snow leopards, red deer, eagles and wild sheep inhabit the high mountain regions, especially in the northern part of the region.

It was through this region of western Mongolia and the Altai Mountains that Genghis Khan entered Mongolia in the twelfth century. Originally named Temujin, the fierce warrior united the Mongol tribes into one nation. As emperor of Mongolia, he created one of the largest and most powerful empires in the history of the world.

In 1206, his fierce warriors conquered the vast area in Asia, which became known as the Mongol Empire. It was the largest adjoining land empire in the history of the world.

The emphasis on riding horses, as an integral part of Mongolian cultures, has been traced to the Bronze Age, around 3,000 BC. Horses thrive in the higher altitudes. Many species inhabit the region's Tavan Bogd National Park. In this region, Belukha Mountain is the highest region of this spectacular mountain range, with its peak reaching nearly 15,000 feet. The highest elevations are those mountains along Mongolia's border with Russia.

There are over thirty glaciers in this region. The ancient occupants also left engraved petroglyphs in this rugged and beautiful region of glaciers and lakes.

In some regions, year-round ice and snow cover the mountain peaks. Glaciers glisten in the greatest heights. Below the spectacular mountains, many wild foxes and bears roam freely. In the warmer months, herds of livestock in the region roam over the rich pasture land. In the short summer months, clear mountain lakes flourish.

Terelj National Park is today home to nomadic tribes who have tended their herds in these pastures for centuries.

In the southwest, on Mongolia's border with China, the mountainous region is called the "Gobi Altai". These mountains tower over the Gobi Desert in the southernmost part of Mongolia.

This region is also known for its fine craftsmen who excel in tapestries and carpets with intricate designs.

Valley near Altai Mountains

Horse Ranch near Altai Mountains

Morning on the Ranch

Ger with a View

Rich grasslands of the Altai Valley

Horses Play a Vital Part in Ranch Life in Mongolia

Young Rider at Work on the Ranch.

Horses on an Early Morning Run

Young Rider Demonstrates his Skills

Young Rider Works with his Favorite Horse.

Early morning training exercises.

The mountains tower above the small
village of gers in the valley below.

Young Mother at Home in her Ger

Stove Inside the Young Mother's Ger

Sleeping Area Inside the Home

Baby Inside the Ger

Traditional Items Inside the Ger

A Line of Gers Across the Valley

Some Gers are Rented Out to Travelers

Inside a Tourist Ger

Horses on a Farm

Morning on the Farm

Young Children in Front of their Home

Gers Overlooking the Valley

Woman Working Inside her Ger

Woman Preparing Food Inside her Ger

Woman Resting Next to her Stove

Village of Gers Dwarfed by the Mountains

An Afternoon Ride through the Valley

Forest and Mountains in the Altai Region

An Eagle, Trained by "Eagle Hunters", to Catch their Prey

Massive Rock Formations in the Altai Mountain Region

Sunny Afternoon in the Altai Mountains

Line of Gers at the foot of the Mountains

Ritual Prayer Flags

Closeup View of Prayer Flags

Backpacking Tourists Heading for the Mountains

Hikers Dwarfed by the High Mountains

Closer View of Hikers Navigating in the Mountains

Inscriptions on the Mountains

CHAPTER 4

Central Mongolia

Ulaanbaatar, Capitol of Mongolia

The city of Ulaanbaatar, located in Central Mongolia, is the country's largest city. It is also the capital of Mongolia. Ulaanbaatar is situated in the middle of Mongolia in a region called Tuv Aimag. It is the center for Mongolia's commerce and trade.

In stark contrast to the serene and sweeping landscapes of the countryside, this city is a bustling, thriving metropolis. Today, the city has a population of over a million residents. Nearby is an expansive National Park called "Bogd Khan Uul". Just outside the city are walking trails to the lush countryside, and stunning mountains rising in the distance.

While some spectacular modern buildings now dominate the center of the city, there are still some older districts on the outskirts of the city. In these nearby neighborhoods, some poverty still exists. Nevertheless, both rich and poor co-exist in peace in this most remarkable of cities.

Founded in 1639 as a Buddhist monastery, Ulaanbaatar is located in the Turel River Valley. Nomadic tribes gravitated here in the 1700's and established a larger Buddhist community. Just outside the city, animals graze freely in the expansive fields.

In the 1900's, Russia and other countries took an interest in this area, primarily for opportunities in the region's fur trading. Foreigners were also interested in profiting from the region's coal mines and gold. Russia invaded the territory and killed religious adherents. The shrines were desecrated and ransacked. Much later, those houses of Buddhist worship were used to house public offices and museums.

Over time, an eclectic group of neighborhoods grew. Such districts as Zaisan, Rapid Harsh Town, and Blue Sky Town eventually coalesced on the Mongolian plain. They would later converge to become the city of Ulaanbaatar. Overall, there are nine municipal areas. These districts, called "duuregs", are represented in the Mongolian parliament by the elected assembly of representatives from the respective districts. These form the "Great Khural State"---the political and legislative body of Mongolia.

Most buildings and infrastructure in Ulaanbaatar were built by the Russians during Soviet rule. When Mongolia declared independence in 1921, Ulaanbaatar became the capital. This city was located on the primary ancient trade routes. Also, the railway system connected the chief trading posts in both China and Russia with Mongolia's rail system.

Over one-third of Mongolia's people live in Ulaanbaatar. It struggles with crime and the normal challenges of large cities.

The city, itself, is in a valley, but its elevation is still over 4,000 feet high. There are four mountain peaks on the outskirts of the city.

The "Blue Sky" building was one of the first ultra-modern buildings on the city's skyline. Its still distinctive outline casts an impressive silhouette against the city's skyline.

Here the old meets the new. There is a synergy about the way they co-exist in a peaceful merging of "what has been" and "what is yet to be."

The old trade route extended from Beijing in China, across Mongolia, and across the Russian border. The two main trading partners for Ulaanbaatar were the city of Kalgan in China, and the city of Maimaicheng in Russia. Today, the main railway is still the primary transport route. The two main trading destinations are also the same.

Ulaanbaatar is considered to be the coldest city, not only in Mongolia, but also in the world. It is located at an altitude of almost 5,000 feet above sea level. During January, its coldest month, the city temperatures are well below any other city, registering temperatures as low as - 16C.

Despite the challenging weather, half of Mongolia's three million people live in this city.

In Mongolia there is really only one city to speak of—Ulaanbaatar. At the cultural heart of the country there is a battle going on—maintaining the old way of life or surrendering to the new. There is a lot at stake.

In this city, there is an undercurrent of energy that is palpable. It is here that the old meets the new. Here sublime sculptures stand next to areas of dilapidated structures, where some locals visibly struggle to exist. Also, near the glistening public buildings with slick white exteriors, there are pockets of poverty which lie in their shadow.

For centuries, the native Mongolians have lived a Nomadic way of life sustaining themselves and the land. Today, the young people head toward the cities for education and for jobs with the blessing of their families. And yet the infrastructure does not yet exist to support them all.

The old ways will not endure, and the new is not yet prepared for them. The young generation—cut from their nomadic roots and way of life—are set adrift. It is a dilemma I have seen nowhere else so sharply as here—the old ways and the new colliding, and in deadlock, despite everyone's efforts and best intentions.

Older Mongolians try to hold, intact, the land and the culture of their fathers in circumstances that are now difficult and not easily sustainable. Alcoholism is a major problem in both the cities and in the rural nomadic culture that struggles to survive. Even with an education, this does not guarantee jobs, and the young men and women who have turned their backs on the past, even with the blessing of their parents, do not yet have an infrastructure that can provide meaningful lives in the cities where unemployment remains high.

In spite of this, it is impossible not to be overwhelmed with the warmth of the native people who cling to the land, who attempt to hold on to both the old and the new, and who are so gracious to strangers.

Life can be very difficult in the cities, especially for newcomers. This is not intended as a criticism. It is simply the face of real life.

This is the place where people from the countryside have come with their dreams. And that is what makes it so important.

The people are the treasure here. For many, city life would seem to offer a way out. And certainly, there are opportunities which abound here. The museums are excellent. There are music venues and theaters which offer the newest productions. The people are friendly and accommodating.

The central square in Mongolia's capitol is the Sukhbaatar Square. It is named for the revolutionary fighter Damdin Sukhbaatar, who declared Mongolia's independence from China in July of 1921. He died two years later in 1923.

In 2006, Sukhbbaatar Square was completed to celebrate the Coronation of Khan over 800 years earlier. Among many of the local highlights are the Government Palace, the Choijin Lama Museum, the State Opera and Ballet Center, and the Gandan Monastery.

Some of the architecture is extraordinary. Tucked in among the public buildings are some quiet public squares with tourists and locals feeding pigeons. There are parks and gardens where locals and tourists gather.

Ulaanbaatar also has some superb ballet and opera venues, and also an excellent orchestra. Both contemporary and traditional theaters can be found in this city.

In some places, there are small gers, like the traditional gers in the countryside. Here, however, some of them are crowded together behind makeshift fences.

It is an interesting mix. There are active night-life spots with restaurants and shops. But most of all, the hospitality and warmth of the people make this an eclectic and exciting place, which should not be missed.

Sports events are also found here in Ulaanbaatar, particularly wrestling which is a favorite national sport. The wrestling events take place in this city during the Nadam Festival.

Sometimes city residents go to the picturesque Orkhon Valley, although it is some distance from the city. There are also biking trails near Ureeg Nuur Lake.

The Migjid Janraisig Sum Temple in Ulaanbaatar

An important landmark in the city is the Migjid Janraisig Sum Temple in Ulaanbaatar. It is part of the Gandan Monastery, the chief monastery in all of Mongolia. This Buddhist temple is a spectacular example of Tibetan Buddhism. Pine trees line the street as you approach the temple.

Inside the Monastery, it is like standing in a massive ancient cathedral. The site is 6 kilometers west of Sukhbaatar Square. The grounds are the property of the Gandantegchinlen Monastery. The monastery is usually referred to simply as "The Monastery of Gandan". This is located in the Bayangol district of the city.

Early in the morning, the voices of chanting monks fills the vast space. Gandan is the main monastery in Mongolia. The name "Janraisig" is the "Bodhisatva", which refers to the deity who is the "Protector of Humans and Animals." Migjid Janraisig Sum is also called "the Bodhisattva of Compassion".

This deity is depicted with four arms. In his four hands, he holds different objects. The items include a vessel, a mirror, a scarf and a pair of spheres. The elongated earlobes on the statue represent compassion and divine wisdom.

This current statue, which replaced the original statue, dates from 1996. Depicted to the right of this statue is Tara, his consort.

The temple has the distinction of having the world's tallest indoor statue. It stands 26.5 meters. This is equivalent to a two-story building in height. Gold leaf has been applied to the entire statue, and over 2,000 precious stones are inlaid in the gold surface.

There was an earlier statue, nearly identical, which the Russians stole from this monastery. This occurred during the Soviet era. The statue which previously stood here was 6 meters higher. It had been erected at the beginning of the twentieth century. Shortly before World War II, it was destroyed by Russian soldiers.

The National Museum of Mongolia

The National Museum of Mongolia opened in 1971. This remarkable museum addresses every aspect of Mongolian life through art, ceremonial costumes, statues and scientific collections relating to Mongolian culture.

The many exhibits range from pre-historic artifacts to modern art.

The Genghis Khan Equestrian Monument

There is a remarkable monument in Ulaanbaatar dedicated to Genghis Khan. It is known as the "Genghis Khan Equestrian Statue". It was erected in 2008 by an architect named Enkhjargal and a renowned sculptor names Erdenebileg. The enormous undertaking was a daring engineering challenge.

From a distance it is a small speck glittering on the vast plateau. There were are no towns or shops nearby. Slowly the massive statue comes into view with the massive Genghis Khan sitting atop his horse. Local legend says that this spot was chosen because it is the place where Genghis Khan found the famous "whip", which became a symbol of his quest to conquer the world, and inspired him to build his kingdom.

In the 13th Century, the rugged explorer had conquered half of the known world.

He had come, however, from a difficult childhood. Born in 1162, his father had been poisoned by a rival tribe, and the family was thrust into poverty. At the age of nine, he was betrothed to a girl named Borte, and was sent to live with her family, until their marriage when he was 16. Meanwhile, he had briefly returned home, and killed his half-brother who was intending to force his own mother to marry him.

Genghis Khan had been born into a turbulent era. From the time of the first Emperor of China in 215 BC, the Chinese had been usurping the nomadic lands, and threating the Mongolian nomads as uncivilized. For over a thousand years, rivalries erupted and lives were lost. The objective of Genghis Khan was to consolidate the nomadic tribes on the Mongolian plateau.

He built a magnificent palace at Karakorum in the thirteenth century, and began to conquer most of the known world. His palace was elaborate with four gigantic gates facing in four directions. Nearby was the Erdene Zuu Monastery, one of the oldest in Mongolia.

When his descendent, Kublai Khan, came to power, he relocated the capital to Dadu.

Unlike his contemporaries who would mutilate and torture their conquests, Genghis Khan spared those who did not resist. Those who resisted, however, met instant death. And the casualties of his armies were significant. In the building of his empire, there were an estimated 40 million who died during the decades of war.

About twenty miles east of Ulaanbaatar, Mongolia's capital, the imposing statue rises out of a remote and isolated field. The great warrior Genghis Khan, posed astride his magnificent horse, gleams in the morning sun. The enormous statue sits atop a four million dollar visitor complex. The statue faces east, looking toward the warrior's birthplace.

This is the world's largest equestrian statue. It was made from 250 tons of steel, and it gleams in the sunlight. In this modern and unexpected construction, visitors can walk up inside the statue, and look out from the head of the horse. They can also walk outside to a viewing platform to appreciate the enormity of the statue, and the sweeping view of the surrounding landscape.

It is a rare and extraordinary experience. There is also an underground museum at this site. The Visitor Center, itself, is 32 feet high. Inside, it also houses a Museum of the Bronze Age.

The Genghis Khan Monument is not only a prestigious tribute to the great leader. It also paid tribute to his religious tolerance. One remarkable thing was that during his reign, he was known for his tolerance of Buddhist monasteries,

Christian churches and Islamic mosques, which flourished side by side in his kingdom.

Amid the raw clash of ancient kingdoms, where brutality was the law, he was known to have been more humane in victory, in many respects, than most of his contemporaries. In the brutal siege for power, Genghis allowed those who surrendered to live and to be assimilated into his empire. Some of his fiercest enemies, whom he had defeated in battle, were to later become his finest generals. Under his rule of law, he took the conquered tribes under his protection, and they became part of his tribe. Even orphans from the conquered tribes were adopted— some by his own family.

He initiated a Code of Law called the Yassa Code, which laid out the policies of war and regulated civilian life. Under his code, he outlawed the kidnapping of women by rival tribes. This was partly in response to the taking of his own young wife when he was in his mid-teens, whom he subsequently rescued. He declared all children to be legitimate, and condemned the practice of selling women into marriage. He made the taking of lost property and the stealing of animals a capital offense. He also regulated hunting laws for the benefit of the people. He revived and regulated trade along the Silk Road which prospered both Europe and Asia. These were progressive ideas for the twelfth century.

His Code of Law, applying to the logistics of war, ordered that those who surrendered were to be spare violence, and he condemned the abuse of people by his troops. Once when a ruler in Baghdad gifted him with a regiment of Crusaders from Europe who had been prisoners, Genghis Khan freed them, and sent them home.

Most importantly, he declared freedom of religion throughout his empire. In his own army Buddhists, Christians and Muslims fought side by side.

Ultimately, Genghis Khan was a brilliant leader who founded the world's largest empire. In 1949, the Soviet Union attempted to diminish his legacy, claiming that he was brutal. The fact is that Genghis Khan was a powerful warrior whose military skills created the largest empire the world has ever known. That empire covered 12 million square miles. It included parts of what are now Russia, Armenia, Iraq, Iran, Afghanistan, Mongolia, Tibet, China, Korea and Eastern Europe. The name "Genghis Khan" means literally, "Universal Ruler".

The total territory of the land he conquered comprised over twenty percent of the earth. Estimates suggest that his subjects numbered a hundred million inhabitants of this massive empire.

Since foreign occupation in the 1920's, there has been a concerted effort to downplay the achievements of Genghis Khan. In 1990, Mongolians rejected Soviet interference. Then in 2008, the people rallied around their Mongolian hero, and revived his memory with the magnificent statue of Genghis Khan.

In a modern-day DNA inquiry, it was determined that a Y-chromosomal link, identified with Genghis Khan, indicated that one in every two hundred men alive today is related to the ancient warrior.

The Genghis Khan Monument was an epic achievement. At his death, Genghis Khan told his three sons, *"With Heaven 's aid I have conquered for you a huge empire. But my life was too short to achieve the conquest of the world. That task is left for you."*

Ulaanbaatar, Capital of Mongolia

The Statue of Buddha overlooks new construction in Ulaanbaatar,
as the old sections of the city surrender to the new.

The signs of a transformation into a modern city in Ulaanbaatar began with daring new architecture.

The "Blue Sky Building" was one of the earliest of the new era buildings in Ulaanbaatar, with its graceful architecture.

New architecture contrasts well with the city's original buildings.

New architecture contrasts well with the city's original buildings.

Young Girls in a Ballet Class in the City

Young Boy Using Transit System in the City

Water Ride in a City Park

Statue in the City of Ulaanbaatar

Popular Night Life with Musical Performers

Female Musicians Perform

Night performance by Female Dancer

Young Boy at School

There are still some pockets of poverty.
Gradually these problem areas are being addressed.

New neighborhoods will replace poorer districts.

Migjid Janraisig
Sum Temple

Inside the Gandan Monastery in the Bayangol District. This
is the most important monastery in all of Mongolia.

This statue of Tara, the Bodhisattva of Compassion, is on display in the Gandan Monastery. It is one of the world's tallest indoor statues.

Exquisite craftsmanship is shown in this elaborate
statue of a deity inside Gandan Monastery.

This is the Stupa outside Gandan Monastery.

This is a small courtyard near Gandan Monastery.

*National Museum
of Mongolia*

The National Museum of Mongolia

Inside the National Museum of Mongolia, this
photo displays several Buddhist artifacts.

Elaborate costumes are showcased inside the National Museum of Mongolia. They date to several different time periods.

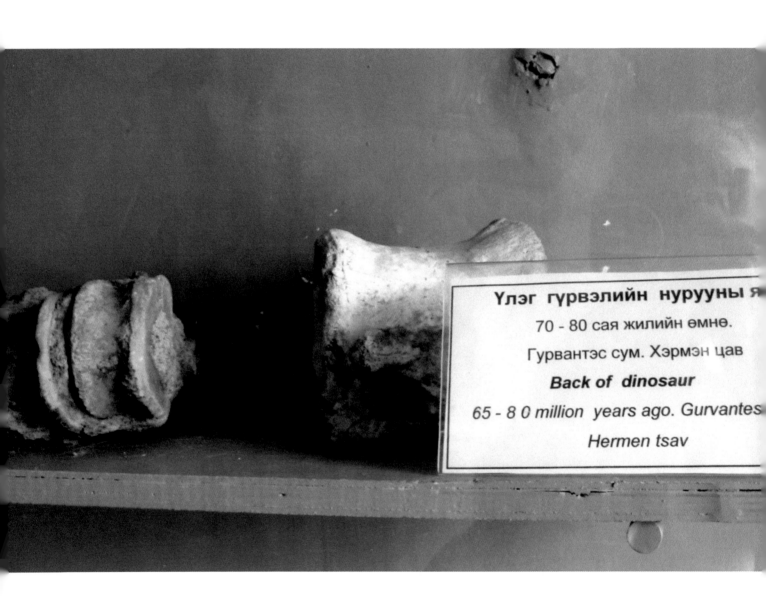

Үлэг гүрвэлийн нурууны я

70 - 80 сая жилийн өмнө.

Гурвантэс сум. Хэрмэн цав

Back of dinosaur

65 - 8 0 million years ago. Gurvantes

Hermen tsav

This display shows dinosaur bones from a dinosaur
spine, dating back over 65 million years.

Үлэг гүрвэлийн хавирганы яс

65- 70 сая жилийн өмнө.

Гурвантэс сум. Нэмэгт уул

Rib of Dinosaur

65 - 70 million years ago. Gurvantes sum.
Nemegt mountain

This display shows dinosaur bones from the rib of
a dinosaur, dating back over 65 million years.

Prehistoric Cave Art

Elaborate Horse Saddle on Display

Ancient Rock Fragment on Display

Miscellaneous Artifacts on Display

Traditional Mongolian Costumes

Female Deity in Bronze

Courtyard near the National Museum of Mongolia

Genghis Khan Monument
near Ulaanbaatar

Face of Genghis Khan Statue

Genghis Khan on his Horse Atop the Monument Museum.

Closeup of the Hand of Genghis Khan on his Statue

Closeup of the Horse's Legs on the Genghis Khan Statue

Statues of Genghis Leading His Warriors
at the Entrance to the Monument.

Printed in the United States
by Baker & Taylor Publisher Services